Preface

There are many good books. But it's difficult to get interesting teaching materials simply written. I'm planning a new series of school textbooks. It has the following features

- written in a clear and concise
- The example contains concept and application
- Bilingual in English and Chinese

Do each question you self   This is the best way to learn

I hope this book would give much help to the readers

Shi Kwok Wong

序

好書很多. 然而要尋得簡明有趣的教材也決非易事

我已考慮寫一套新的系列教學參考書　它應具有下述

特色:

- 簡明易懂

- 例題包涵概念和運用

- 英中雙語參考

動手做每一題　這才是最好的學習方法

我希望　本書對讀者能有所幫助

　　　　　　　　　　　王士國

Contents

1   Number system   ………………… ………… . .5

1-1 Fundamental system…………………………   ….5

1-2 Represent   ………………………………..   5

1-3 Symbol   …………………………. .   5

1-4 Definition   …………………………   6

2   Decimal system   ………………………… …   8

3   Binary system   ………………………… ..   11

4   Hexadecimal system   ……………………..   14

5   Conversion between number system…   …. 17

6   Application   …………………………………   20

目錄

1 數字系統 28

1-1 基本系統 28

1-2 表示方法 28

1-3 符號 28

1-4 定義 29

2 十進制系統 31

3 二進制系統 34

4 十六進制系統 37

5 數字系統之轉換 40

6 運用 ....43.............50

1   NUMBER SYSTEM

1-1   Fundamental system

Decimal system

Binary system

Hexadecimal system

1-2   Represent

1   Decimal system

→   (decimal number)$_d$

$_d$ (decimal number.)   may be omitted

2 Binary system

→ (Binary number)$_b$

3 Hexadecimal system

→ (hexadecimal number)$_h$

1-3   Symbol

1   Decimal number

0,   1,   2,   3,   4,   5,   6,   7,   8,   9.

2   Binary number

(0,   1 .)$_b$

Note $0_b = 0$ $1_b = 1$.

3.   Hexadecimal number

(0, 1, 2, 3, 4, 5, 6, 7, 8, 9, A, B, C, D, E, F.)$_h$

Note

$0_h = 0$	$1_h = 1$	$2_h = 2$	$3_h = 3$
$4_h = 4$	$5_h = 5$	$6_h = 6$	$7_h = 7$
$8_h = 8$	$9_h = 9$	$A_h = 10$	$B_h = 11$
$C_h = 12$	$D_h = 13$	$E_h = 14$	$F_h = 15$

1-4   Definition

1   Place value P.V.

The place of the digit '1' equals decimal

value, called the Place Value

2   Digit place value D.P.V.

= Digit $\times$ P.V.

= D $\times$ P. V.

Note   The same digit D

D $\times$ left P. V. $>$ D $\times$ right P. V.

The Same P.V.

If   D↑,   D↑ × P.V.   then   D.P.V.↑

What conclusion we can get now ?

Example

4 5 6   and   7 8 9

P.V. of 6 and 9 is 1

P.V. of 5 and 8 ia 10

P.V. of 4 and 7 is 100

But

  D.P.V. of 6 is 6

  D.P.V. of 9 is 9

  D.P.V. of 5 is 50

  D.P.V. of 8 is 80

  D.P.V. of 4 is 400

  D.P.V. of 7 is 700

Understand ?

2   DECIMAL SYSTEM

Base - 10

0,   1,   2,   3,   4,   5,   6,   7,   8,   9.

Conclusion

1. Minimum   0

2. Maximum   9   i.e.   10 - 1.

3   Each digit   <   Base - 10   i.e. < 10

4.   All digit   10

5.   Place value   P.V. = $(10^{n-1})_d$

N   →   4,   3,   2,   1,   0,   -1,   -2, ...

P.V. → ··· 10^3, 10^2, 10^1, 10^0, $\frac{1}{10^1}$, $\frac{1}{10^2}$, $\frac{1}{10^3}$ ···

        i.e. ...   100, 10,   1, $\frac{1}{10}$, $\frac{1}{100}$,...

Neighbor $\frac{10^{n-1}}{10^{n-2}}$ = 10

i.e. $10^{n-1} = 10 \times 10^{n-2}$

i.e.   Left P.V. = 10 × Right P. V.

i.e.   Left P.V.   >   Right P.V.

6   A number $\times$ 10^n $\rightarrow$ point moves n place

If   n > 0 $\bullet$ $\rightarrow$

If   n < 0 $\leftarrow$ $\bullet$

Example

$655.36 \times 10^1 = 6553.6$ $\bullet$ $\rightarrow$

$655.36 \times 10^2 = 65536.$ $\bullet$ $\rightarrow$

$655.36 \times 10^{-1} = 65.536$ $\leftarrow$ $\bullet$

$655.36 \times 10^{-2} = 6.5536$ $\leftarrow$ $\bullet$

7.   Addition in the same P.V.

If the sum ≥ 10 carry 1 to the left

Example

$$\begin{array}{r} \bullet\ \ 1 \\ +\ \ \ 9 \\ \hline 1\ \ \ 0 \end{array} \qquad \begin{array}{r} \bullet\ \ 2 \\ +\ \ \ 9 \\ \hline 1\ \ \ 1 \end{array}$$

8.   Subtraction in the same P.V.

Borrowing 1 from left equals 10

Example

$$\begin{array}{r} \bullet\ \ 1 \\ -\ \ \ 9 \\ \hline 2 \end{array} \qquad \begin{array}{r} \bullet\ \ 2 \\ -\ \ \ 9 \\ \hline 3 \end{array}$$

9. Multiplication

$$128 \times 2$$

$$= 128 \times (1 + 1)$$

$$= 128 + 128$$

$$= 256$$

We are changing multiplication into addition.

10. Division

$$256 \div 128 = 2$$

$$(128 \times 2 = 256)$$

$$256 - 128 = 128$$

$$128 - 128 = 0$$

we are changing division into subtraction

.

3   BINARY SYSTEM

Base - 2

0,   1.

Conclusion

1.   Min   0

2.   Max   1   i.e.   2 - 1

3   Each digit   <   Base - 2   i.e. <   2.

4.   All digit   2

5   Place value   P.V. = $(2^{n-1})_d$

N $\rightarrow$ 4,   3,   2,   1,   0,   -1,   -2,.....

P.V.$\rightarrow$...   , 2^3 , 2^2 , 2^1 , 2^0 , $\frac{1}{2^1}$, $\frac{1}{2^2}$, $\frac{1}{2^3}$ ....

...8 ,   4 ,   2 ,   1 , $\frac{1}{2}$, $\frac{1}{4}$, $\frac{1}{8}$ ...

Neighbor $\frac{2^{n-1}}{2^{n-2}} = 2$

i.e. $2^{n-1} = 2 \times 2^{n-2}$

i.e. Left P.V. = 2 $\times$ Right P.V.

i.e. Left P.V. > Right P.V.

6. A number$_b$ $\times$ 2^n $\rightarrow$ The point moves n place

If   n > 0   • $\rightarrow$

If   n < 0 $\leftarrow$ •

Example

$(1\ 11.11)_2 \times 2^1$

$= (1\ 111.1)_2$ $\bullet$ $\rightarrow$

$(1\ 11.11)_2 \times 2^2$

$= (1\ 1111)_2$ $\bullet$ $\rightarrow$

$(1\ 11.11)_2 \times 2^{-1}$

$= (1\ 1.111)_2$ $\leftarrow$ $\bullet$

$(1\ 11.11)_2 \times 2^{-2}$

$= (1.1111)_2$ $\leftarrow$ $\bullet$

7   Addition in the same P.V.

If   Sum = 2   Carry 1 to the left

Example

```
      •   1
   +      1
   ───────────
      1   0
```

8   Subtraction in the same P.V.

Borrowing 1 from left equals 2.

Example

```
      •   0
   −      1
   ───────────
          1
```

9   Multiplication

$$1001_b \times 11_b$$

$$= 1001_b \times (1 + 1 + 1)_d$$

$$= (1001 + 1001 + 1001)_b$$

$$= 11011_b$$

Change multiplication into addition

10   Division

$$(11011 \div 1001)_b = 11_b$$

$$(11011 - 1001 - 1001 - 1001)_b$$

$$= (10010 - 1001 - 1001)_b$$

$$.= (1001 - 1001)_b$$

$$= 0$$

Change division into subtraction

4   HEXADECIMAL SYSTEM

Base - 16

0, 1, 2, 3, 4, 5, 6, 7, 8, 9, A, B, C, D, E, F.

Conclusion

1. Min   0

2   Max   F   i.e. 16 - 1

3   Each digit   <   Base - 16   i.e. < 16

4   All digit   16

5   Place value   P.V. = $(16^{n-1})_d$

n $\rightarrow \cdots$ 4,   3,   2,   1,   0,   -1,   -2...

P.V. $\rightarrow$ ... $16^3, 16^2, 16^1, 16^0, \frac{1}{16}, \frac{1}{16^2}$, ..

i.e. ...   ... 256 , 16 ,   1, $\frac{1}{16}, \frac{1}{256}$, ...

Neighbor $\frac{16^{n-1}}{16^{n-2}} = 16$

i.e. $16^{n-1} = 16 \times 16^{n-2}$

i.e.   Left P.V. = 16 × Right P.V.

i.e.   Left P.V. > Right P.V.

6   A number $_h \times 16^n$ The point moves n place

   If $n > 0$ •   →

   If $n < 0$ ←   •

Example

$100.00_h \times 16^1 = 1000_h$

$100.00_h \times 16^2 = 10000_h$

$100.00_h \times 16^{-1} = 10_h$

$100.00_h \times 16^{-2} = 1_h$

7   Addition in the same P.V.

If   Sum ≥ 16 Carry 1 to the left

Example

```
  •    1        •    2
+      F      +      F
1      0      1      1
```

8   Subtraction in the same P.V.

   Borrowing 1 from left equals 16

Example

```
  •    0
−      F
       1
```

9   Multiplication

$A5_h \times 3$

$= (A5 + A5 + A5)_h$

$= (14A + A5)_h$

$= 1EF_h$

Change multiplication into addition

10   Division

$1EF_h \div A5_h = 3$

$(1EF - A5 - A5 - A5)_h$

$= (14A - A5 - A5)_h$

$= (A5 - A5)_h$

$= 0$

Change division into subtraction

5 CONVERSION BETWEEN NUMBER

- d → b

$$2^{n-1} \dots 2^4 \quad 2^3 \quad 2^2 \quad 2^1 \quad 2^0 \quad 2^{-1} \quad 2^{-2} \quad 2^{-3} \quad 2^{-4} \dots$$

P.V....16   8   4   2   1 $\frac{1}{2}$ $\frac{1}{4}$ $\frac{1}{8}$ $\frac{1}{16}$ ...

10

Example

convert 10 into number $_b$?

We can see that number 10 is a number between P.V. 16 and P.V. 8

So   8 + 2 = 10   number $_b$ = 1010 $_b$

- d → h

$$16^{n-1} \dots 16^3 \quad 16^2 \quad 16^1 \quad 16^0 \quad 16^{-1} \quad 16^{-2} \dots$$

P.V ...   4096   256   16   1 $\frac{1}{16}$ $\frac{1}{256}$ ...

100

Example

Convert 100 into number $_h$?

Number 100 between P.V. 256 and 16

$16 \times 8 = 128 > 100$

$16 \times 7 = 112 > 100$

$16 \times 6 = 96 < 100$

So $\quad 6 \times 16 + 4 \times 1 = 64_h$

- b $\rightarrow$ d

Find the sum of each digit D.P.V.

Example

$(10001)_b$

$= 1 \times 16 + 0 \times 8 + 0 \times 4 + 0 \times 2 + 1 \times 1$

$= 17$

- b $\rightarrow$ h

Group the digits in sets of four, beginning at the right

Example

$(1100001100010010)_b$

$= (1100 \quad 0011 \quad 0001 \quad 0010)_b$

$= (C312)_h$

- h → d

Find The sum of all D.P.V.

Example

$C312_h$

$C \times 16^3 + 3 \times 16^2 + 1 \times 16 + 2 \times 1$

$= 12 \times 16^3 + 3 \times 16^2 + 1 \times 16 + 2 \times 1$

$= 49938$

- h → b

Change each place into binary

$(C312)_h$

$= (\quad C \qquad 3 \qquad 1 \qquad 2 \quad)_h$

$= (\ 1100 \quad 0011 \quad 0001 \quad 0010\ )_b$

6 APPLICATION

1. If 40_a = 60_b find the minimum a and b.

A and b are the positive integers

   Solution

In the number system

4 < a   6 < b

P.V. $...a^2$ +   a   +   1   + ...

$...b^2$ +   b   +   1   +...

So $4 \times a + 0 \times 1 = 6 \times b + 0 \times 1$

4 a = 6 b

Method 1.   See my work (H.C.F. and L.C.M.)

         a   →   12,   15 ,   18...

                           b   →   8,   10,   12 ...

Mini a = 12   Mini b = 8

Method 2   find   4 a = 6 b

4 < a   6 < b

$4 \times (1,   2,   3, ...12 ....)$

$6 \times (1, \quad 2, \quad 3, \quad \ldots .. \ 8 \ldots \ldots)$

The result is the same

Mini a = 12   Base = 12

Mini b = 8   Base = 8

2   Find the maximum and minimum decimal number. Using each of the digits

$(0,2,5,7)_8$ Once only

Solution   See P7

Maxi $(7520)_8$

Mini $(0257)_8$

$7 \times 8^3 + 5 \times 8^2 + 2 \times 8 + 0 \times 1 = 3920$

$2 \times 8^2 + 5 \times 8 + 7 \times 1 = 175$

So   Maxi decimal is 3920

    Mini decimal is 175

3   A, B, C, D, E, F, G, H respectively represents a decimal digit of 0, 1, 2, 3, 4, 5,

6, 7, 8, 9. Find the addend and the sum of the following addition operation

$$\begin{array}{r} H\ D\ E\ F\ F \\ +\ B\ A\ G\ D\ C \\ \hline B\ A\ E\ D\ C\ H \end{array} \quad \ldots (1)$$

Solution

If we know the value of the capital notation also know the value of addend and sum

Let $\quad C_6, C_5, C_4, C_3, C_2, \rightarrow$ Carry

$$
\begin{array}{r}
C_6\,C_5\,C_4\,C_3\,C_2 \\
H\;D\;E\;F\;F \\
+\quad B\;A\;G\;D\;C \\
\hline
B\;A\;E\;D\;C\;H
\end{array}
\quad\ldots\quad (1)
$$

$B = C_6$

If $\quad C_6 = 0 \qquad C_5 = 0$

$(1)\ldots\quad
\begin{array}{r}
0\;0 \\
HDEFF \\
+\quad 0AGDC \\
\hline
0AEDCH
\end{array}
\quad$ SO $\quad H = A \quad$ No

If $\quad C_6 = 0 \qquad C_5 = 1$

$(1)\ldots\quad
\begin{array}{r}
0\,1\,c_4 \\
HDEFF \\
+\quad 0AGDC \\
\hline
0AEDCH
\end{array}$

$\qquad C_4 + D + A = C_4 + D + (1 + H) = 10 + E$

Or $\quad 10^3 (C_4 + D + A)$

$\qquad = 10^3 [C + D + (1 + H)] = 10^4 + 10^3 E$

A equation – four unknown $\qquad\qquad$ No

If $\quad C_6 = 1 \qquad C_5 = 0$

$(1)\ldots\quad
\begin{array}{r}
1\,0 \\
HDEFF \\
+\quad 1AGDC \\
\hline
1AEDCH
\end{array}$

$0 + H + 1 = 10 + A \quad$ So $\quad H = 9 \quad A = 0 \quad$ Ok

(1)..
$$\begin{array}{r} {\scriptstyle 10} \\ 9DEFF \\ +\ 10GDC \\ \hline 10EDC9 \end{array}$$
..... (2)

Consider

$C_4 = 0$

$D + 0 = E \quad D = E \quad$ No

Note: $A = 0 \quad B = 1 \quad H = 9$

$C_4 = 1$

$C_4 + D = 1 + D = E \quad$ Ok

If $C_3 = 0$

$$\begin{array}{r} E \\ +\ G \\ \hline D \end{array} \rightarrow \begin{array}{r} D \\ 1 \\ +\ G \\ \hline D \end{array}$$

 i.e. $D + 1 + G = 10 + D \quad G = 9 \quad$ No

If $C_3 = 1,$
$$\begin{array}{r} 1 \\ E \\ G \\ \hline D \end{array} \rightarrow \begin{array}{r} 1 \\ D \\ 1 \\ +\ G \\ \hline D \end{array}$$

i.e. $D + 2 + G = 10 + D \quad G = 8$

from (2)
$$\begin{array}{r} {\scriptstyle 1011} \\ 9DEFF \\ +\ 108DC \\ \hline 10EDC9 \end{array}$$
...(3)

 note $0 = A \quad 1 = B \quad 9 = H$

If $D = 2 \quad F = 3, 4, 5, 6, 7. \quad C_3 = 0 \quad$ No

If   D = 3   F = 2, 4, 5, 6. $C_3 = 0$ No

           F = 7 $C_3 = 1$ but   C = 0   No

If   D = 4   F = 2, 3, 5. $C_3 = 0$ No

           F = 6 $C_3 = 1$ but   C = 0   No

           F = 7 $C_3 = 1$ but   C = 1   No

If   D = 5,   F = 2, 3, 4. $C_3 = 0$ No

           F = 6 $C_3 = 1$ but   C = 1   No

           F = 7 $C_3 = 1$ C = 2   Ok

So   D = 5   F = 7   C = 2   E = D + 1 = 5 + 1 = 6

From (3)
$$\begin{array}{r} 1011 \\ 9DEFF \\ +\underline{108DC} \\ 10EDC9 \end{array} \quad \rightarrow \quad \begin{array}{r} 95677 \\ +\underline{10852} \\ 106529 \end{array}$$

So   Addend   HDEFF = 95677

             BAGDC = 10852

   Sum   BAEDCH = 106529

4   Compare conversion resources

For example   certain number (99999999)

9 9 9 9 9 9 9 9 $\rightarrow$ 8-place lengths

= 5F5E0FF$_h$ $\rightarrow$ 7-place lengths

=101 1111 0101 1110 0000 1111 1111

$\rightarrow$ 27-place lengths

What conclusion we can get?

For example   certain length (4 place)

Maximum decimal 9999

Maximum binary   1111$_b$ $\rightarrow$ 15

Maximum hexadecimal FFFF$_h$ $\rightarrow$ 65535

What conclusion?

5   Convert fraction into binary

- P.V. conversion

Example $8\frac{7}{9} \rightarrow b$

P.C.   …4, 2, 1, 1/2, 1/4, 1/8, …

So $8\frac{7}{9} = 1000.110001_b$

- The point conversion

Equal conversion

Example $\frac{3}{4} = \frac{11_b}{4} = 0.11_b$

I.e. the point moves two places to the left

Note   3 $\rightarrow$ integer

4   $\rightarrow$ integer. P.V. = 2^2

Unequal conversion

Example

$\frac{3}{5} = \frac{11_b}{5} \approx 0.10011 ..._b$

i.e.   the point moves two places to the left ....

Note   3 $\rightarrow$ integer

5 $\rightarrow$ integer   P.V. $\neq$ 5

Different number system

- Base – 8   0, 1, 2,3, 4, 5, ,6, 7.

P.V. ... 8^3, 8^2, 8^1, 8^0, 8^{-1}, 8^{-2}, 8^{-3}, ...

- Base - 12   0, 1, 2, 3, 4, 5, 6, 7, 8, 9, 10,11.

P.V. ... 12^3, 12^2, 12^1, 12^0, 12^{-1}, 12^{-2}, 12^{-3}...

- Base - 24   0,1,2,3,............20,21,22,23.

P.V... 24^3, 24^2, 24^1, 24^0, 24^{-1}, 24^{-2}, 24^{-3}...

- Base - 60   0,1,2,3,4,......................57,58,59.

P.V. ...60^3, 60^2, 60^1, 60^0, 60^{-1}, 60^{-2}, 60^{-3},..

- Base - ? ...

1 數字系統

1-1 基本系統

十進制系統

二進制系統

十六進制系統

1-2 表示方法

1 十進系統

（十進數）$_d$

→十進數 $_d$ 可不寫

2 二進系統

→（二進數）$_b$

3 十六進系統

→ (十六進數)$_h$

1-3 符號

1. 十進數

0, 1, 2, 3, 4, 5, 6, 7, 8, 9.

2. 二進數

(0, 1.)$_b$

注意 $0_b = 0$ $1_b = 1$

3 十六進數

(0, 1, 2, 3, 4, 5, 6, 7, 8, 9, A, B, C, D, E, F.)$_h$

注意

$0_h = 0$　　　$1_h = 1$　　$2_h = 2$　　$3_h = 3$

$4_h = 4$　　　$5_h = 5$　　$6_h = 6$　　$7_h = 7$

$8_h = 8$　　　$9_h = 9$　　$A_h = 10$　　$B_h = 11$

$C_h = 12$　　$D_h = 13$　　$E_h = 14$　　$F_h = 15$

1-4 定義

1 位值 P.V.

• 數字 '1' 所在位置 等值的十進制數

名為 位值

2 數字位值 D.P.V.

= 數字 × P.V. = D × P.V.

注意　　同一數 D

　　　　D × 左項 P.V. > D × 右項 P.V.

　　　　同一 P.V.

如　 D↑　則 D↑× P.V.　則 D.P.V.↑

由此　有何結論 ？

例

　456　和　789

6 和 9 的　位值是 1

5 和 8 的　位值是 10

4 和 7 的　位值是 100

但

6 的　數字位值　是 6

9 的　數字位值　是 9

5 的　數字位值　是 50

8 的　數字位值　是 80

4 的　數字位值　是 400

7 的　數字位值　是 700

明白嗎?

2　　十進制系統

基 - 10

0,　1,　2,　3,　4,　5,　6,　7,　8,　9.

推論

1　最小數字　　0

2　最大數字　　9　即 10 —1.

3　每一數字　<　基數 - 10　　即 <　10

4　所有數字　　10 個

5　位值　　P.V. = $(10^{n-1})_d$

N → 　.... 　4,　3,　2,　1,　0,　-1,　-2, ...

P.V. → ··· 　$10^3, 10^2, 10^1, 10^0, \frac{1}{10^1}, \frac{1}{10^2}, \frac{1}{10^3}$ ···

　　　即　　...100, 10,　1,　$\frac{1}{10}$,　$\frac{1}{100}$, ...

相鄰兩項　$\frac{10^{n-1}}{10^{n-2}}$ = 10

即　$10^{n-1} = 10 \times 10^{n-2}$

即　左項 P.V. = 10 × 右項 P.V.

即　左項 P.V. > 右項 P.V.

6　一數 × 10^n → 小數點 移動 n 位

　如　n > 0　　•　→

如　n　<　0　　←　•

例

$655.36 \times 10^1 = 6553.6$　•　→

$655.36 \times 10^2 = 65536.$　•　→

$655.36 \times 10^{-1} = 65.536$　←　•

$655.36 \times 10^{-2} = 6.5536$　←　•

7.　同位值加法

如果　和 ≥ 10　向左進位 1

例

$$
\begin{array}{r} \bullet\ 1 \\ +\ 9 \\ \hline 1\ 0 \end{array} \qquad
\begin{array}{r} \bullet\ 2 \\ +\ 9 \\ \hline 1\ 1 \end{array}
$$

8　同位值減法

向左借 1　等　10

例

$$
\begin{array}{r} \bullet\ 1 \\ -\ 9 \\ \hline 2 \end{array} \qquad
\begin{array}{r} \bullet\ 2 \\ -\ 9 \\ \hline 3 \end{array}
$$

9　乘法

128×2

$$= 128 \times (1 + 1)$$

$$= 128 + 128$$

$$= 256$$

以上 將乘法運算 變為 加法運算

10 除法

$$256 \div 128 = 2$$

$$(128 \times 2 = 256)$$

$$256 - 128 = 128$$

$$128 - 128 = 0$$

以上 將除法運算 變為 減法運算

.

3　二進制系統

基 - 2

0,　1

推論

1　最小數字　0

2　最大數字　1　即 2 - 1

3　每一數字 <　基數 - 2　即　< 2.

4　所有數字　2 個

5　位值 P.V. = $(2^{n-1})_d$

N→ ....　　4,　3,　2,　1,　0,　-1,　-2,.....

P.V.→...　　2^3　2^2　2^1　2^0　$\frac{1}{2^1}$,　$\frac{1}{2^2}$,　$\frac{1}{2^3}$ ....

即:　8,　4,　2,　1,　$\frac{1}{2}$,　$\frac{1}{4}$,　$\frac{1}{8}$, ..

相鄰兩項　$\frac{2^{n-1}}{2^{n-2}} = 2$

即　$2^{n-1} = 2 \times 2^{n-2}$

即　左項 P.V. = 2 × 右項 P.V.

即　左項 P.V. >　右項 P.V.

6　一數$_b \times 2^n$　→　小數點移 n 位

如　n > 0　•　→

如　n < 0　←　　•

例

$(1\,11.11)_2 \times 2^1$

$= (1\,111.1)_2$　　•　→

$(1\,11.11)_2 \times 2^2$

$=(1\,1111)_2$　　•　　→

$(1\,11.11)_2 \times 2^{-1}$

$= (1\,1.111)_2$　　←　•

$(1\,11.11)_2 \times 2^{-2}$

$= (1.1111)_2$　　←　•

7　同位值加法

如　和 = 2　向左進位 1

例

```
•   1
+   1
1   0
```

8　同位值減法

向左借 1 等 2

例

$$\cdot \quad 0$$
$$=\ \underline{\quad 1}$$
$$\qquad 1$$

9   乘法:

$$1001_2 \times 11_2$$

$$= 1001_2 \times (1+1+1)_{10}$$

$$= (1001 + 1001 + 1001)_2$$

$$= 11011_2$$

以上 將乘法運算 變為 加法運算

10   除法

$$(11011 \div 1001)_2 = 11_2$$

$$(11011 - 1001 - 1001 - 1001)_2$$

$$= (10010 - 1001 - 1001)_2$$

$$= (1001 - 1001)_2$$

$$= 0$$

以上 將除法運算 變為 減法運算

4 十六進制系統

基 - 16

0, 1, 2, 3, 4, 5, 6, 7, 8, 9, A, B, C, D, E, F.

推論

1 最小數字 0

2 最大數字 F 即 16 - 1

3 每一數字 < 基 - 16 即 < 16

4 所有數字 16 個

5 位值 P.V. = $(16^{n-1})_d$

$n \rightarrow \cdots, \quad 4, \quad 3, \quad 2, \quad 1, \quad 0, \quad -1, \quad -2\ldots$

$P.V. \rightarrow \ldots 16^3, \quad 16^2, 16^1, 16^0, \frac{1}{16}, \frac{1}{16^2}, \frac{1}{16^3}, \cdots$

即 ..256, 16, 1, $\frac{1}{16}$, $\frac{1}{256}$, ..

2 相鄰兩項 $\frac{16^{n-1}}{16^{n-2}} = 16$

即 $16^{n-1} = 16 \times 16^{n-2}$

即 左項 P.V. = 16 × 右項 P.V.

即 左項 P.V. > 右項 P.V.

3 一數 $_h$ × 16^n 小數點 移 n 位

如 n > 0 • →

如　n < 0　　← ·

例

$100.00_h \times 16^1 = 1000_h$

$100.00_h \times 16^2 = 10000_h$

$100.00_h \times 16^{-1} = 10_h$

$100.00_h \times 16^{-2} = 1_h$

7　同位值加法

如　和　≥ 16　向左進 1

例

```
  ·   1          ·   2
+   F          +   F
─────          ─────
1   0          1   1
```

8　同位值減法

向左借 1 等 16

例

```
  ·   0
−   F
─────
    1
```

9　乘法

$A5_h \times 3$

$= (A5 + A5 + A5)_h$

$= (14A + A5)_h$

$= 1EF_h$

以上 將乘法運算 變為 加法運算

10　除法

$1EF_h \div A5_h = 3$

$(1EF - A5 - A5 - A5)_h$

$= (14A - A5 - A5)_h$

$= (A5 - A5)_h$

$= 0$

以上 將除法運算 變為 減法運算

5 數字系統之轉換

• d → b

| 2^{n-1} | ... | 2^4 | 2^3 | 2^2 | 2^1 | 2^0 | 2^{-1} | 2^{-2} | 2^{-3} | 2^{-4} | ... |

P.V....16　8　4　2　1　$\frac{1}{2}$　$\frac{1}{4}$　$\frac{1}{8}$　$\frac{1}{16}$...

10

例

10　→　二進數 $_b$?

可以看到 10 位於 P.V.16 和 P.V.8.之間

所以　8 + 2 = 10　二進數 $_b$ = 1010_b

• d → h

| 16^{n-1} | | 16^3 | 16^2 | 16 | 16^0 | 16^{-1} | 16^{-2}... |

P.V　4096　256　16　1　$\frac{1}{16}$　$\frac{1}{256}$...

100

例

100　→　十六進數 $_h$?

100 位於 P.V.256 和 P.V.16. 之間

$16 \times 8 = 128 > 100$

$16 \times 7 = 112 > 100$

$16 \times 6 = 96 < 100$

所以 $6 \times 16 + 4 \times 1 = 64_h$

- b → d

求出 每一個數字的 數字位值 之和 (即可)

例

$(10001)_b$

$= 1 \times 16 + 0 \times 8 + 0 \times 4 + 0 \times 2 + 1 \times 1$

$= 17$

- b → h

四個一組, 從右開始

例

$(1100001100010010)_b$

$= (1100 \quad 0011 \quad 0001 \quad 0010)_b$

$= (\quad C \quad 3 \quad 1 \quad 2 \quad)_h$

$= (C312)_h$

- h → d

求出 所有數字位值 之和

例

C312$_h$

$= C \times 16^3 + 3 \times 16^2 + 1 \times 16 + 2 \times 1$

$= 12 \times 16^3 + 3 \times 16^2 + 1 \times 16 + 2 \times 1$

$= 49938$

- h → b

將 每一個十六進數 轉換為 四個二進數

例

(C312)$_h$

$= ($ C   3   1   2 $)_h$

$= ($ 1100 0011   0001 0010$)_b$

6　運用

6-1　如 $40_a = 60_b$. 求 a 和 b 的最小值. a 和 b　　　為正整數

解

在數字系統中　基 > 每一個數字

即　　$4 < a,$　　　　$6 < b$

P.V.　$...a^2 + a + 1 + ...$

$...b^2 + b + 1 + ...$

所以　　$4 \times a + 0 \times 1 = 6 \times b + 0 \times 1$

$4a = 6b$

方法 1.　參看我所著（H.C.F.和 L.C.M.）一書

　　　　　　a　→　12,　15,　18...

b　→　8,　10,　12 ...

最小數 a = 1 2　最小數 b = 8

方法 2　　求 $4a = 6b$

$4 < a$　　$6 < b$

$4 \times (1,$　　$2,$　　　$3, ... 12, ...)$

$6 \times (1,$　　$2,$　　　$3, ...8, ...)$

結果相同

最小數 = 12　　基 = 12

最小數 b = 8　　基 = 8

6-2 求最小十進數. 每一數字 (0,2,5,7)$_8$ 僅取一次

求最大十進數. 每一數字 (0.2.5.7)$_8$ 僅取一次

解

　參看 P38

最大數　　(7520)$_8$

最小數　　　(0257)$_8$

$7 \times 8^3 + 5 \times 8^2 + 2 \times 8 + 0 \times 1 = 3920$

$2 \times 8^2 + 5 \times 8 + 7 \times 1 = 175$

所以　最大十進數　3920

　　最小十進數　175

6-3　　A, B, C, D, E, F, G, H 分別表示一個十進數 0, 1, 2, 3, 4, 5, 6, 7, 8, 9. 求下

述 加法運算中的 加數及和

$$\begin{array}{r} H\ D\ E\ F\ F \\ +\ B\ A\ G\ D\ C \\ \hline B\ A\ E\ D\ C\ H \end{array} \quad ....(1)$$

解

如知道 字母符號之值 則加數及和 即可知之

考慮　C_6, C_5, C_4, C_3, C_2 表示 ，→ 進位

$$\begin{array}{r} c_6\,c_5\,c_4\,c_3\,c_2 \\ H\,D\,E\,F\,F \\ +\,B\,A\,G\,D\,C \\ \hline B\,A\,E\,D\,C\,H \end{array} \quad \dots \quad (1)$$

$B = C_6$

如　$C_6 = 0$　　$C_5 = 0$

$$(1)\ \dots +\begin{array}{r} 00 \\ H\,D\,E\,F\,F \\ 0\,A\,G\,D\,C \\ \hline 0\,A\,E\,D\,C\,H \end{array} \quad 所以\ H = A\ 不可$$

如　$C_6 = 0$　$C_5 = 1$

$$(1)\dots +\begin{array}{r} 01c_4 \\ H\,D\,E\,F\,F \\ 0\,A\,G\,D\,C \\ \hline 0\,A\,E\,D\,C\,H \end{array}$$

$C_4 + D + A = C_4 + D + (1 + H) = 10 + E$

或　$10^3 (C_4 + D + A)$

$= 10^3 [C_4 + D + (1 + A)] = 10^4 + 10^3 E$

一個等式　四個未知數　不可

如　$C_6 = 1$　$C_5 = 0$

$$(1)\dots +\begin{array}{r} 10 \\ H\,D\,E\,F\,F \\ 1\,A\,G\,D\,C \\ \hline 1\,A\,E\,D\,C\,H \end{array}$$

$0 + H + 1 = 10 + A$　　所以　$A = 0$　$H = 9$　　可以

$$(1)\dots\begin{array}{r} 10 \\ 9\,D\,E\,F\,F \\ +\quad 1\,0\,G\,D\,C \\ \hline 1\,0\,E\,D\,C\,9 \end{array} \quad \dots (2)$$

考慮

$C_4 = 0$

$D + 0 = E$ $D = E$ 不可

注意 $A = 0$ $B = 1$ $H = 9$

$C_4 = 1$

$C_4 + D = 1 + D = E$ 可

如 $C_3 = 0$

$$\begin{array}{r} E \\ + G \\ \hline D \end{array} \quad \rightarrow \quad \begin{array}{r} D \\ 1 \\ + G \\ \hline D \end{array}$$

即 $D + 1 + G = 10 + D$ $G = 9$ 不可

如 $C_3 = 1$ $\begin{array}{r} 1 \\ E \\ + G \\ \hline D \end{array} \quad \rightarrow \quad \begin{array}{r} 1 \\ 1 \\ D \\ + G \\ \hline D \end{array}$

即. $D + 2 + G = 10 + D$ $G = 8$

從（2） $\begin{array}{r} 1011 \\ 9DEFF \\ + \quad 108DC \\ \hline 10EDC9 \end{array}$ ....(3)

注意 $0 = A$ $1 = B$ $9 = H$

如 $D = 2$ $F = 3, 4, 5, 6, 7.$ $C_3 = 0$ 不可

如 $D = 3$ $F = 2, 4, 5, 6.$ $C_3 = 0$ 不可

          $F = 7$ $C_3 = 1$ 但 $C = 0$ 不可

如　D = 4　　F = 2, 3, 5.　　　　$C_3 = 0$　　不可

　　　　　　　F = 6　$C_3 = 1$　　但　C = 0　　不可

　　　　　　　F = 7　　$C_3 = 1$　　但　C = 1　　不可

如　D = 5　　F = 2, 3, 4.　　　　$C_3 = 0$　　不可

　　　　　　F = 6　　$C_3 = 1$　但　C = 1　　不可

　　　　　　　　F = 7　　$C_3 = 1$　　C = 2　　　可

所以　D = 5　　F = 7　C = 2　E = D + 1 = 5 + 1 = 6

從 (3)　$\quad \begin{array}{r} 1011 \\ 9DEFF \\ + \quad 108DC \\ \hline 10EDC9 \end{array} \quad \rightarrow \quad \begin{array}{r} 95677 \\ + 10852 \\ \hline 106529 \end{array}$

所以　　加數　　HDEFF = 95677

　　　　　　　　BAGDC = 10852

　　　和　　　BAEDCH = 106529

6-4　　比較轉換資源

例　定數 (9999 9999)

9999　9999　→　8 - 位字長

= 5F5E0FF$_h$　　→　7 - 位字長

=101 1111 0101 1110 0000 1111 1111

→ 27 - 位字長

有何推論 ？

例　定長 (4 位)

　最大　9999

　最大　1111_b　→　15

　最大　$FFFF_h$　→　65535

有何推論 ？

6-5　分數轉 二進數

- 　位值 P.V. 法

：　$8\frac{7}{9}$　→　b

P.C.　...4,　2,　1,　1/2,　1/4,　1/8, ...

所以　$8\frac{7}{9} = 1000.110001_b$

- 小數點法

等值轉換

例

$$\frac{3}{4} = \frac{11_b}{4} = 0.11_b$$

即小數點 向左移 2 位

注意 3 → 整數

4 → 整數. P.V. = 2^2

不等轉換

例

$$\frac{3}{5} = \frac{11_b}{5} \approx 0.10011\ldots_b$$

即小數點向 左移 2 位....

注意 3 → 整數

5 → 整數 P.V. ≠ 5

6-6 不同的數字系統

- 基 - 8 0, 1, 2, 3, 4, 5, ,6, 7.

P.V. ... 8^3, 8^2, 8^1, 8^0, 8^{-1}, 8^{-2}, 8^{-3}, ...

- 基 - 12 0, 1, 2, 3, 4, 5, 6, 7, 8, 9, 10,11.

P.V. ... 12^3, 12^2, 12^1, 12^0, 12^{-1}, 12^{-2}, 12^{-3}...

- 基 -24 0,1,2,3,............20,21,22,23.

P.V... 24^3, 24^2, 24^1, 24^0, 24^{-1}, 24^{-2}, 24^{-3}...

- 基 - 60 0, 1, 2, 3,.............57, 58, 59.

P.V. ...60^3, 60^2, 60^1, 60^0, 60^{-1}, 60^{-2}, 60^{-3}...

- 基 - ? ...